green my
experience

SUSTAINABILITY DICTIONARY

A pocket guide to basic sustainability terms designed to accelerate your sustainability journey.

THE
LITTLE
GREEN
BOOK

green**my**
experience

[The Little Green Book - Sustainability Dictionary]

[www.greenmyexperience.com]

[© Green My Experience, 2024   ]

Print ISBN: 979-8-9905861-1-6

Ebook ISBN: 979-8-9905861-0-9

green my
experience

This **Sustainability Dictionary - The Little Green Book,** provides you with an orientation to basic sustai- nability terms and concepts. Designed for both the newbie and the expert, these simple definitions will help dee-pen your own understanding of sustainability and help you teach others about critical sustainability concepts.

At **Green My Experience**, we align with the mindset that sustainability is not just about the environment and climate, but rather encompasses environmental, social, and governance factors or "ESG" (see how we snuck your first definition in here?). Check us out at www.greenmyexperience.com

Terms explored in this guide:

Definitions to take first steps
on the path to sustainability.

green^{my}
experience

START HERE: BASIC DEFINITIONS YOU NEED TO KNOW

☑ **Basic Definitions:**
1. Sustainability
2. Sustainability Strategy
3. Sustainable Development Goals (SDGs) SDGs
4. Environmental, Governance, & Social (ESG)
5. Triple Bottom Line (People, profit, planet)
6. Sustainability Report
7. Stakeholder
8. Materiality
9. Certification

☑ **Enviromental Definitions:**
10. Eco-Efficiency
11. Circular Economy
12. Greenhouse Gases (GHG)
13. Greenhouse Gas Inventory
14. Carbon Footprint
15. Mitigation Strategies
16. Adaptation
17. Carbon Net Zero
18. Carbon Positive
19. Natural Capital

☑ **Social Definitions:**
20. Human Rights
21. Worker's Rights
22. Occupational Health & Safety
23. Gender Equality
24. Diversity
25. Social Investment
26. Community Engagement
27. Disability Rights

☑ **Governance Definitions:**
28. Ethics Code
29. Due Diligence
30. Sustainable Consumption
31. Sustainable Value Chain

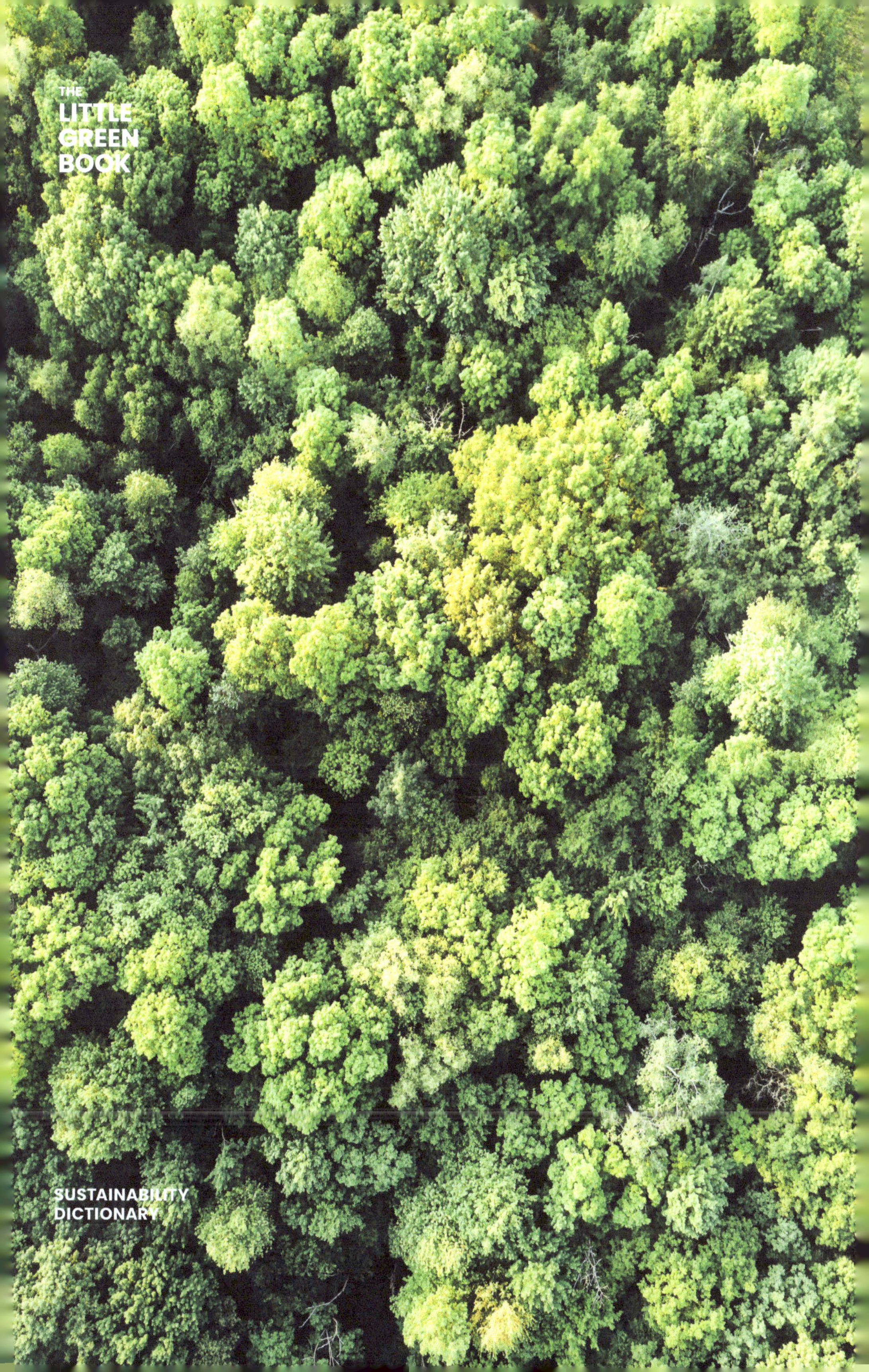

THE LITTLE GREEN BOOK

SUSTAINABILITY DICTIONARY

green my experience

BASIC DEFINITIONS:

1. Sustainability

2. Sustainability Strategy

3. Sustainable Development Goals (SDGs)

4. Environmental, Governance, & Social (ESG)

5. Triple Bottom Line (People, profit, planet)

6. Sustainability Report

7. Stakeholder

8. Materiality

9. Certification

1. SUSTAINABILITY

Sustainability means doing things now in a way that doesn't make it hard for people in the future to do what they need to do. It's like making sure you have enough cookies for yourself but also saving some for your friends who will come over later. The United Nations Brundtland Commission said sustainability is about meeting our needs today without stopping the next generation from meeting their needs. This includes thinking about nature, people, and money, so that we can have a better and fairer world in the future.

2. SUSTAINABILITY STRATEGY

A detailed plan created by an organization to tackle their environmental, social, and economic challenges in a way that's good for the company, its people, and the environment. It includes goals, ideas, and rules aimed at making sure the organization company acts responsibly and fairly.

3. SUSTAINABLE DEVELOPMENT GOALS (SDGs)

The Sustainable Development Goals (SDGs) are a set of 17 goals created by the United Nations in 2015 as part of Agenda 2030. These goals cover important areas like education, gender equality, clean energy, economic growth, climate action, ecosystem conservation, and peace. They are all connected because improving one area often helps improve others.

The SDGs are like a roadmap for countries, businesses, and groups to work together to make the world better for everyone. They guide cooperation among governments, companies, and communities to make sure we're moving towards a more sustainable future and a better quality of life for people all around the world.

4. ENVIRONMENTAL, GOVERNANCE, & SOCIAL(ESG)

ESG stands for Environment, Social, Governance. While these are three separate terms, they are often grouped together as a way for organizations to think about their sustainability strategy.

Environment: This is about how a company affects the planet, like its carbon footprint or how it uses natural resources.

Social: This is about how a company treats people, like its employees, customers, and communities where it operates.

Governance: This is about how a company is managed and controlled, including things like transparency, ethics, and accountability.

ESG helps companies and investors consider not only financial performance but also their impact on the environment, society, and how they're run. It's becoming increasingly important as businesses strive to be more responsible and sustainable.

5.TRIPLE BOTTOM LINE (People, Profit, Planet)

Triple Bottom Line (people, profit, planet) is a way of running a business that cares about three main things: making money (profit), treating people well (people), and taking care of the environment (planet). It's about measuring success not just by how much money is made, but also by how happy people are and how little harm is done to the planet.

Some well-known companies who are known for focusing on this approach include:

Patagonia: A clothing company that emphasizes sustainability and environmental responsibility in its business practices.

Ben & Jerry's: An ice cream company known for its social activism and commitment to environmental sustainability.

These companies are recognized for prioritizing not only profits but also the well-being of people and the planet in their business strategies.

6. SUSTAINABILITY REPORT

A sustainability report is a document that shows how well a company or organization is doing in terms of being sustainable. Typically produced annually to show year-over-year progress, sustainability reports provide transparency to investors, customers, and the general public about an organization's progress towards their sustai- nability goals and their key ESG metrics. This report helps people see if the company is taking care of the planet and its people while also making money.

7. STAKEHOLDER

A stakeholder is anyone who could be influenced by what a company or organization does, or who might care about its decisions. This includes customers, employees, investors, the board, and the communities where the business operates. It can also include environmental factors like the land and wildlife affected by the company's actions.

8. MATERIALITY

Materiality refers to figuring out which issues really matter to stakeholders when they judge how well a company is doing, whether it's about money, the environment, or social matters. For example, for a beverage company like Coca-Cola, water scarcity is a big deal. They need water to make their drinks, and if they can't get enough water or if it becomes too expensive, it could cause big problems for their business. Also, since others in the community need water too, Coca-Cola might face pressure from people who are concerned about where they get their water from.

9. CERTIFICATION

Certification means meeting specific standards or rules that show a company or product is doing well in areas like the environment, social responsibility, or sustainability. These standards are usually set by an outside group and checked by a third party to make sure they're being followed. It's like getting a stamp of approval to show that a company or product is doing the right things. For instance, B Corp, ISO standards, and LEED are examples of certifications that companies can get to prove they're doing good things for the planet and society.

SUSTAINABILITY
DICTIONARY

greenmy
experience

ENVIRONMENTAL DEFINITIONS:

10. Eco-efficiency

11. Circular Economy

12. Greenhouse Gases (GHG)

13. Greenhouse Gas Inventory

14. Carbon Footprint

15. Mitigation Strategies

16. Adaptation

17. Carbon Net Zero

18. Carbon Positive

19. Natural Capital

10. ECO-EFFICIENCY

Eco-efficiency is a business method that focuses on using resources wisely and reducing harm to the environment throughout the entire life of a product or service. With an eco-efficient mindset, companies can find ways to make things like energy, water, and materials last longer and cause less pollution.

Common eco-efficiency practices:
• Reducing the amount of pollution and waste created
• Using less energy and water
• Using more renewable resources
• Encouraging reuse and recycling
• Designing products and processes to be more eco-friendly from the start.

11. CIRCULAR ECONOMY

The goal of a circular economy is to keep products and materials in use for as long as possible, cutting down on waste and saving natural resources.

Examples of circular economy practices:
Remanufacture: Fixing up used products to make them like new again. For instance, companies like IKEA with its Take-Back program allow customers to return their furniture to be repurposed.

Waste food reduction: Initiatives to reduce food waste find ways to use surplus food from businesses and households, either by giving it to those in need or turning it into other products like animal feed or compost.

Product as a Service (PaaS): Instead of buying products outright, consumers can subscribe to services that give them access to products when they need them. For example, companies like Philips offer lighting as a service, where customers pay for the service of illumination rather than owning the lighting equipment outright.

12. GREEHOUSE GASES (GHG)

Greenhouse gases are the types of gasses that play a role in global warming and climate change. When these gasses are released, they trap heat in the Earth's atmosphere which leads to higher temperatures on the planet's surface. Some examples of GHGs are carbon dioxide (CO2), methane (CH4), and nitrous oxide (N2O). While these gases occur naturally, human activities like burning fossil fuels and deforestation have made GHG levels go up. This means more heat gets trapped which causes problems for all living things.

13. GREEHOUSE GAS INVENTORY

The Greenhouse gas inventory is a way to measure and report how much greenhouse gas is being produced from activities or sources such as driving cars or running factories. You can even calculate your own greenhouse gas number to see how your daily choices impact the environment. For businesses, collecting this data helps them see how their business is impacting climate change and informs their sustainability strategy.

14. CARBON FOOTPRINT

The total amount of greenhouse gasses released directly or indirectly by a specific entity, event, product, or individual over a set period of time. This footprint is typically measured in terms of carbon dioxide equivalent (CO2e) mass, which includes not only carbon dioxide but also other greenhouse gasses like methane and nitrous oxide.

For instance, the carbon footprint of a product might include emissions from manufacturing, transportation, and disposal. By calcula-ting and understanding our carbon footprint, we can better grasp our impact on the environment and work towards reducing it.

15. MITIGATION STRATEGIES

The practices and policies we put in place to cut down on greenhouse gas emissions, which helps to lower our impact on climate change. We do this by using strategies like leveraging renewable energy, saving energy through reduction strategies, planting trees, and storing carbon. Even personal actions, like using a reusable cup instead of disposable ones is a mitigation strategy that can help lessen our personal contribution to climate change.

16. ADAPTATION

Adaptation is our response to the changes that come with climate change. Some of these changes are unstoppable because of the gasses already building up in the atmosphere. Adaptation helps us limit harm, deal with risks, and make the most of opportunities in our changing world.

17. CARBON NET ZERO

Carbon net zero is the goal to reduce greenhouse gas emissions as much as we can, almost to zero. While there are some emissions that cannot be completely removed, like from factories or cars, those can be balanced out by absorbing or capturing those gasses. For example, this happens naturally when oceans and forests soak up carbon dioxide from the air. The goal is to remove any greenhouse gasses hanging around in the atmosphere.

17. CARBON POSITIVE

Carbon positive is when a company does more than just reduce their emissions to zero via reduction strategies and offsets by taking extra steps to remove more CO_2 from the environment than they put in. For example, The Berkeley Group, UK's first carbon positive company, maintains its carbon positive status by investing in projects that help take carbon dioxide out of the atmosphere. For example, one of their projects, Madre de Dios, reduces deforestation, biodiversity loss, and the threat of illegal logging through forest management practices in the Amazon rainforest. forest management practices in the Amazon rainforest.

18. NATURAL CAPITAL

Natural capital refers to resources from nature that directly or indirectly create value. It includes air, water, soil, minerals, and all the plants and animals (biodiversity) around us. These resources are valuable because they are essential to help us live. For example, natural capital gives us clean air to breathe, fresh water to drink, good soil to grow food, and helps us control the climate, pollinate our crops, and provide places for fun and relaxation. Natural capital is nature's bank account, and it's important for our health, happiness, and economy to keep it a positive one.

SUSTAINABILITY
DICTIONARY

THE
LITTLE
GREEN
BOOK

green my
experience

SOCIAL DEFINITIONS:

20. Human Rights
21. Worker's Rights
22. Occupational Health & Safety
23. Gender Equality
24. Diversity
25. Social Investment
26. Community Engagement
27. Disability Rights

20. HUMAN RIGHTS

Human rights are the basic rights that every person has, no matter where their nationality, place of residence, sex, ethnic origin, color, religion, language, or any other characteristic. [Ideally] these rights are protected by international laws and rules. They include civil, political, economic, social, and cultural rights. Human rights make sure that everyone is treated fairly and has the freedom to live their life without discrimination or harm.

21. WORKER RIGHTS

Workers' rights are a set of rules that protect the rights of people who work. These rules are designed to make sure that workers are treated fairly and safely while on the job. They cover things like fair pay, safe working conditions, and respect. Workers' rights are important to ensure everyone has a safe and positive experience at work, no matter where they are in the world.

22. OCCUPATIONAL HEALTH & SAFETY

Occupational Health & Safety focuses on keeping workers safe and healthy while they're at work, which is one of a worker's rights. It involves creating and following policies to make sure the workplace is free from dangers that could harm employees. This includes providing protective equipment, training workers on safety procedures, and regularly checking for potential hazards. The goal is to minimize any risks to workers' health and safety so they can do their jobs without getting hurt.

23. GENDER EQUALITY

Gender equality means making sure that everyone, no matter the gender, has the same rights, responsibilities, and chances in life. This includes fair access to education, healthcare, and jobs, as well as the freedom to live without being treated unfairly or facing violence because of their gender.

24. DIVERSITY

Diversity refers to the mix of differences and similarities among people. These differences include things like race, ethnicity, gender, age, and more. The concept of diversity recognizes and values the uniqueness of each individual and promotes inclusion and respect for differences in social, educational and workplace settings.

25. SOCIAL INVESTMENT

Social investment is when money, people, and materials are put into programs or projects to help improve the lives of people and communities. This investment aims to make a positive and lasting difference in areas like education, healthcare, housing, jobs, and community development. It's giving support and resources to improve society for everyone, now and in the future.

26. COMMUNITY ENGAGEMENT

Community engagement is when an organization and local communities work together in a positive way. This entails building trust and mutual benefit with residents, community groups, nonprofit organizations, and local authorities.

27. DISABILITY RIGHTS

Disability rights are rules and protections that ensure people with disabilities get to live with the same choices, opportunities, and respect as everyone else. These rights make sure that people with disabilities can go to school, work, and enjoy public places just like people without disabilities and protect them from being treated unfairly because of their disability. This includes making sure buildings have ramps or elevators, providing tools and technology that help disabled persons learn or communicate, and making it illegal to discriminate against someone because they have a disability.

SUSTAINABILITY
DICTIONARY

THE
LITTLE
GREEN
BOOK

greenmy
experience

GOVERNANCE DEFINITIONS:

28. Ethics Code
29. Due Diligence
30. Sustainable Consumption
31. Sustainable Value Chain

28. ETHICS CODE

A set of guidelines that establish standards of behavior and ethics for an organization. In the workplace, codes of ethics provide guidance for decision making and typically address topics like integrity, honesty, respect, fairness, transparency, and social responsibility.

29. DUE DILIGENCE

Due diligence is a detailed review done proactively to spot both the clear and possible negative impacts that an organization's activities could have socially, environmentally, and economically. The goal is to prevent and reduce any harmful effects.

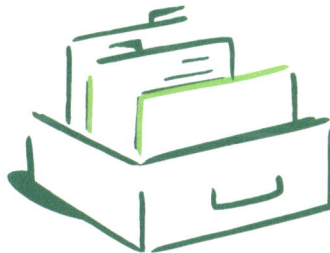

30. SUSTAINABLE CONSUMPTION

Sustainable consumption means using products and services in a way that doesn't harm the environment, society, or economy. It includes taking action to save natural resources, make less waste, use eco-friendly products, and support fair business practices. By making sure products are made responsibly from start to finish, we can help the planet and people while still getting what we need.

31. SUSTAINABLE VALUE CHAIN

A sustainable value chain integrates environmental, social, and economic considerations into all business activities and processes. The goal is to ensure long-term viability of a product or service while protecting the well-being of future generations.

green^my
experience

CREATING A SUSTAINABILITY STRATEGY

How do I get started?

The best way to start is by getting a baseline of your current performance in each area of ESG and begin measuring and tracking your emissions. Once you have a sense of your current state, you can set goals and create strategies to help you meet your sustainability goals. Green My Experience's tools and resources are designed to help you do just that.

About Green My Experience

Green My Experience™ helps organizations and individuals transform to a greener way of being. Through our innovative platform, expert guidance and tools, education, and strategic planning services, we facilitate a deeper connection between nature and organizations while creating gains for people, profit, and planet. Find out more at www.greenmyexperience.com

BIBLIOGRAPHY

Indiana University Environmental Resilience Institute. (s.f.). Greenhouse Gas Inventories. Recuperado de https://eri.iu.edu/tools-and-resources/fact-sheets/greenhouse-gas-inventories.html

Encyclopedia Britannica. (s.f.). Carbon footprint. Recuperado de https://www.britannica.com/science/carbon-footprint

NASA Climate Change. (s.f.). Adaptation & Mitigation. Recuperado de https://climate.nasa.gov/solutions/adaptation-mitigation/#:~:text=Mitigation%20%E2%80%93%20reducing%20climate%20change%20%E2%80%93%20involves,these%20gases%20(such%20as%20the

Net Zero Climate. (s.f.). What is Net Zero? Recuperado de https://netzeroclimate.org/what-is-net-zero-2/

Fast Company. (s.f.). Climate positive, carbon neutral, carbon negative: What do they mean? Recuperado de https://www.fastcompany.com/40583176/climate-positive-carbon-neutral-carbon-negative-what-do-they-mean

Convention on Biological Diversity. (s.f.). Business and Biodiversity: Natural Capital. Recuperado de https://www.cbd.int/business/projects/natcap.shtml

Office of the United Nations High Commissioner for Human Rights. (s.f.). What are human rights? Recuperado de https://www.ohchr.org/en/what-are-human-rights

Global Network for the Right to Food and Nutrition. (s.f.). Workers' Rights. Recuperado de https://globalnaps.org/issue/workers-rights/

International Organization of Employers. (s.f.). Occupational Safety and Health. Recuperado de https://www.ioe-emp.org/policy-priorities/occupational-safety-and-health/

Built In. (s.f.). Diversity & Inclusion. Recuperado de https://builtin.com/diversity-inclusion

Accion Empresas. (2017). INVERSIÓN SOCIAL ESTRATÉGICA Y PRODUCTIVIDAD: Una relación a fomentar. Recuperado de https://accionempresas.cl/content/uploads/documento-inversion-social-1.pdf

Lightning Source LLC
Chambersburg PA
CBHW041715200326
41519CB00001B/176